SAVED
Endangered Species Success Stories

Carla Mooney

ReferencePoint Press

San Diego, CA

ReferencePoint Press®

For more information, contact:
ReferencePoint Press, Inc.
PO Box 27779
San Diego, CA 92198
www.ReferencePointPress.com

LIBRARY OF CONGRESS CATALOGING-IN-PUBLICATION DATA

Name: Mooney, Carla, 1970- author
Title: Saved: Endangered Species Success Stories/by Carla Mooney.
Description: San Diego, CA : ReferencePoint Press, 2027. | Includes bibliographical references and index.
Identifiers: LCCN 2026000581 (print) | LCCN 2026000582 (ebook) | ISBN 9781678213527 library binding | ISBN 9781678213534 ebook
Subjects: LCSH: Endangered species--Juvenile literature | Wildlife conservation--Juvenile literature
Classification: LCC QH75 .M657 2026 (print) | LCC QH75 (ebook)
LC record available at https://lccn.loc.gov/2026000581
LC ebook record available at https://lccn.loc.gov/2026000582

CONTENTS

Saving the Gray Wolf

For at least half a million years, the gray wolf roamed North America. At one point, as many as 2 million gray wolves lived throughout most of the continental United States. When European colonists arrived in North America, they brought cows, sheep, pigs, and other livestock. As livestock grazed near forests and other wild areas, the wolves quickly learned they made easy prey.

To protect their livestock, farmers and landowners hunted and trapped the wolves. Some offered bounties, paying twenty to fifty dollars per wolf kill. Professional wolf hunters trapped, shot, and hunted wolves using dogs. In 1906, the US Forest Service launched efforts to exterminate the wolves. By the 1950s, these extermination efforts had wiped out nearly all the gray wolves in the continental United States. The gray wolf was now in danger of becoming extinct.

Disappearing Species

Around the world, thousands of animal and plant species, like the gray wolf, are at risk of disappearing forever. These species face a variety of threats, from overhunting to habitat loss, that threaten their ability to survive. When one species disappears, entire ecosystems suffer. Animals and plants depend on each other for food, shelter, nutrients, and pollination. The extinction of a single species can create ripple effects throughout an ecosystem, affecting every organism. "Every animal in the ecosystem is important for that ecosystem. They all have a really important role to play,"[1] says Adrienne Crosier, a biologist at the Smithsonian Conservation Biology Institute.

> "Anytime you take a species completely out of the ecosystem, you cause an imbalance in that ecosystem."[2]
>
> —Adrienne Crosier, biologist at the Smithsonian Conservation Biology Institute

In the case of the gray wolf, the loss of wolves significantly impacted the ecosystems where they lived. Without the wolf as a top predator, populations of its prey, such as elk and deer, grew unchecked, leading to overgrazing of native vegetation. The loss of vegetation led to soil erosion, making it more difficult for some plants to grow. With fewer plants available, the birds, animals, insects, and other organisms that relied on them for food and shelter suffered. The ecosystems where the wolf had lived deteriorated. "Anytime you take a species completely out of the ecosystem, you cause an imbalance in that ecosystem,"[2] says Crosier.

Protecting the Wolf

During the late 1960s, attitudes toward the gray wolf began to change. Scientists recognized the interconnectedness of each part of an ecosystem and the gray wolf's role. In 1974, the US Fish and Wildlife Service (FWS) listed gray wolves as endangered under a law known as the Endangered Species Act (ESA). This designation made it illegal to hunt or trap the wolves. At that time, only about one thousand wolves remained in the continental United States. After receiving legal protections, gray wolf populations slowly began to recover in areas such as the Great Lakes, Northern Rockies, and the Pacific Northwest. Some wolves naturally migrated from Canada into the northern United States.

Conservationists also reintroduced the gray wolf to Yellowstone National Park, where the wolf population had disappeared by 1926. In 1995 and 1996, the National Park Service introduced thirty-one Canadian gray wolves into Yellowstone. After their reintroduction, the wolves reestablished themselves in the Yellowstone ecosystem and formed packs, raised pups, and spread into surrounding regions. Yellowstone's ecosystem even became healthier because wolves helped control prey populations, which reduced overgrazing. Currently, there are approximately five hundred wolves in the Greater Yellowstone Ecosystem.

The gray wolf is just one of thousands of animals that is at risk of disappearing due to threats like overhunting and habitat loss. While there was some success in reintroduction into nature, the gray wolf's total recovery is still uncertain.

Habitat protection played an important role in the gray wolf's recovery. National parks, forests, and wilderness areas provided safe places for wolves to live and hunt. As habitats were restored, wolves had more space to thrive. Conservationists also worked to reduce conflicts between wolves and people. Programs were created to help ranchers protect their livestock by using guard animals, fencing, and better herding practices. Some states and organizations also offered compensation programs that paid ranchers for livestock lost to wolves.

Since the 1970s, these conservation efforts have helped gray wolf populations steadily grow. As of 2025, approximately fifty-five hundred wolves live in the continental United States, with another eight to eleven thousand in Alaska, according to the International

Wolf Center. However, the gray wolf's recovery is still uncertain. Multiple attempts to delist the wolf and remove its ESA protections have been made over the years. The most recent action occurred in December 2025, when the US House of Representatives voted to remove the gray wolf's federal protections. If the bill becomes law, the gray wolf will no longer be protected under the ESA, although some state protections may still be in effect.

> **"Wolves made progress towards recovery across the country because of the protections of the Endangered Species Act."[3]**
>
> **—Collette Adkins, carnivore conservation director, Center for Biological Diversity**

Conservationists have spoken out against the action, arguing that the wolf is still vulnerable. "Wolves made progress towards recovery across the country because of the protections of the Endangered Species Act,"[3] says Collette Adkins, the Center for Biological Diversity's carnivore conservation director. She worries that removing protections will cause the wolf population to decline again.

Conservation Saves Species

Like the gray wolf, thousands of species worldwide are threatened and endangered. However, there is hope. The work of many scientists, conservationists, and community organizations is helping many endangered species make a remarkable comeback. These success stories show that saving endangered species is possible. With coordinated conservation efforts, endangered species worldwide can thrive again.

CHAPTER ONE

The State of Endangered Species

The Bornean elephant, a subspecies of the Asian elephant, is the smallest elephant in the world, standing only about 8.2 to 9.8 feet (2.5 to 3.0 m) tall. It has large ears, relatively straight tusks, and a long tail that sometimes touches the ground. The Bornean elephant lives in forests on the island of Borneo. Scientists studying the elephant determined that it was genetically different from other Asian elephants, making it a separate subspecies, *Elephas maximus borneensis*.

These elephants need to roam across large areas to find enough food. Yet that is becoming increasingly difficult. Located in Southeast Asia, Borneo is shared by three countries: Malaysia, Indonesia, and Brunei. On Borneo, the elephants primarily live in the Malaysian state of Sabah, with a small population living in Indonesia's part of the island. Sixty percent of the forests where the elephants live in Sabah have been cut down over the past forty years. Large swaths of forest have been converted into commercial plantations. Logging, small farms, and palm oil plantations have also fragmented and shrunk the elephant's habitat, reducing opportunities for contact among elephant groups. With limited contact, the elephants have fewer opportunities to find mates, which can lead to a loss of genetic diversity.

By 2024, fewer than fifteen hundred Bornean elephants lived on Borneo, according to the World Wildlife Fund (WWF). These animals are a keystone species in Borneo's rainforests, dispersing seeds and creating clearings that support forest regrowth and plant diversity. A keystone species is one that other species depend on in an ecosystem. The loss of this elephant species would have a significant impact on Borneo's ecosystems.

In June 2024, the International Union for the Conservation of Nature (IUCN) officially classified the Bornean elephant as endangered on its Red List of Threatened Species, which provides information about the health of species worldwide. When a species is endangered, its population has dropped so low that it could disappear forever. "As a distinct subspecies of the Asian elephant, these animals are unique to the island of Borneo and part of our natural heritage. The inclusion of Bornean elephants on the Red List is a pivotal step to galvanize conservation efforts and direct resources to areas of utmost importance,"[4] says Cheryl Cheah, an ecologist with the WWF Malaysia.

IUCN Red List Classification

To protect and save species like the Bornean elephant, scientists must first classify their risk of extinction. Making these classifications is the job of the IUCN Species Survival Commission. The commission is made up of more than eight thousand scientists from 162 countries. It periodically reviews the status of every species, including animals, plants, and fungi, with sufficient data to make an assessment for the IUCN Red List.

Commission members receive intense training to evaluate the status of each species. Scientists review data such as population size, number of breeding adults, and geographic range. Using the data, the commission places each species in one of nine categories: not evaluated, data deficient, least concern, near threatened, vulnerable, endangered, critically endangered, extinct in the wild, and extinct. Sometimes a species has not been evaluated, or more data are needed, and its classification reflects this. Species that are classified as being of least concern and near threatened are not at immediate risk of extinction. All species classified as vulnerable, endangered, or critically endangered are considered to be threatened, which means they face the risk of extinction.

Logging, farming, and oil plantations are shrinking the habitat of the Bornean elephant and causing a loss of genetic diversity in this species.

Species in the final two categories are already extinct either in the wild or entirely.

A species that lives in different ranges can be placed in different Red List classifications. As a result, a species can be classified as being of least concern in one region but endangered in another. For example, gray whales living in the western North Pacific are considered critically endangered on the IUCN Red List due to significant threats from oil and gas development and fishing, but a healthy population of gray whales lives in the eastern Pacific Ocean along the North and South American coasts.

> **"Species around the world are under huge pressure. So no matter where you look, the numbers of threatened species are rising."[5]**
>
> —Craig Hilton-Taylor, head of the IUCN Red List unit

As of December 2025, the IUCN Red List classified more than 48,600 species, including animals, plants, and fungi, as threatened with extinction. That number is approximately 28 percent of all species assessed by the commission. "Species around the world are under huge pressure. So no matter where you look, the numbers of threatened species are rising,"[5] says Craig Hilton-Taylor, head of the IUCN's Red List unit.

The Endangered Species Act

Although the Red List identifies species at risk, the IUCN lacks legal authority to require anyone to protect those species. However, governments and nongovernmental organizations worldwide use the IUCN Red List to guide their decisions regarding conservation efforts and environmental policies. In the United States, the 1973 ESA protects plants and animals in danger of disappearing. Under the ESA, species can be classified as either endangered or threatened, depending on how close they are to extinction. "Threatened species by definition are at an inflection point—they are not yet in danger of extinction, but are likely to become so in the foreseeable future if nothing is done to reverse

The International Union for Conservation of Nature

The International Union for Conservation of Nature (IUCN) is a global organization based in Switzerland that works to protect nature and reduce biodiversity loss. Founded in 1948, the IUCN is the world's oldest global environmental organization. It brings together governments, scientists, nonprofit organizations, and local communities to find solutions to environmental challenges. One of its most important contributions is the IUCN Red List of Threatened Species, which is a comprehensive information source that assesses the conservation status of plants and animals worldwide. These rankings help governments and conservation groups decide where to focus protection efforts. The categories are as follows:

- Extinct: No individuals remain anywhere on Earth.
- Extinct in the wild: Survives only in captivity or as a naturalized population outside its historic range.
- Critically endangered: Faces an extremely high risk of extinction in the wild.
- Endangered: Faces a very high risk of extinction in the wild.
- Vulnerable: Faces a high risk of extinction in the wild.
- Near threatened: Close to qualifying for a threatened category or likely to do so soon.
- Least concern: Widespread and abundant; currently at low risk.
- Data deficient: Not enough information to assess extinction risk.
- Not evaluated: Has not yet been assessed against the criteria.

> "Threatened species by definition are at an inflection point—they are not yet in danger of extinction, but are likely to become so in the foreseeable future if nothing is done to reverse their trajectory."[6]
>
> —Ellen Richmond, senior attorney with Defenders of Wildlife

their trajectory,"[6] says Ellen Richmond, senior attorney with Defenders of Wildlife, a nonprofit conservation organization. Each category comes with specific protections. When a species is added to the list, geographical areas important to its survival are identified. Certain activities in these areas may be limited to prevent harm to the species. In most cases, it is illegal to kill, harm, or disturb listed animals unless a special permit is granted. Violating the law can result in large fines or even jail time.

To decide which species are protected under the ESA, the FWS and the National Oceanic and Atmospheric Administration (NOAA) National Marine Fisheries Service consider information from several sources, including the IUCN Red List. Individuals can also submit a petition for these agencies to list a species under the act. To be considered for listing, there must be evidence that a species is threatened or endangered because of a loss of habitat or range, overexploitation, disease or predation, natural or human threats, or the failure of other regulations or measures to protect a species. FWS and NOAA scientists review each petition and decide whether to classify a particular species under the ESA as endangered or threatened. Today, nearly twenty-four hundred animal and plant species are protected under the ESA, with almost seventeen hundred of those species located within the United States.

Since its passage, the ESA appears to be working. "The ESA is the cornerstone of wildlife protection in the U.S. Since its enactment more than 50 years ago, the ESA has been remarkably effective at protecting our nation's biodiversity: almost every listed species is still with us today. Hundreds are on the path of recovery,"[7] says Richmond.

Threats to Endangered Species

Endangered species face many serious threats, mostly caused by human activities. Habitat loss is the biggest threat; forests, wetlands, and grasslands are destroyed as people create farms, expand cities, and build roads. Animals and plants need specific places to live, find food, and raise their young. The orangutan is a species that has struggled because of habitat loss and destruction. Bornean orangutans live in Borneo's rainforests. However, these forests are being cut down to make way for palm oil plantations and other human development. As a result, only about 104,700 Bornean orangutans live in the wild, and the IUCN has listed the species as critically endangered.

Poaching and overharvesting threaten species by reducing populations faster than they can reproduce. Poaching is the illegal hunting or capturing of animals, often for fur, horns, tusks, or the wildlife trade for exotic pets. Overharvesting can happen even when hunting or fishing is legal but occurs too often or in large numbers. When species are taken faster than they can reproduce, their populations decline. Fish populations, for example, can collapse when too many fish of a particular species are caught before they reach breeding age. Once numbers drop too low, it can be very difficult for a species to recover.

Intense wildfires and other extreme weather events caused by climate change are destroying the habitats of many species.

Extinct: The Passenger Pigeon

The passenger pigeon is an example of what can happen if a species is not protected. The passenger pigeon was once one of the most abundant bird species in North America, with billions of birds flying across the skies in large flocks. These birds lived in large forests and ate tree nuts, such as acorns. During the 1800s, many people hunted passenger pigeons for food, feathers, and profit. Because the birds traveled in large groups, they were easy targets for hunters. At the same time, deforestation destroyed much of their natural habitat. As forests disappeared and hunting continued without restrictions, the passenger pigeon population declined rapidly. The last known wild passenger pigeon was shot in 1900. The world's last surviving passenger pigeon, a bird named Martha, lived in captivity at the Cincinnati Zoo and died in 1914. With her death, the species disappeared from Earth.

Climate change is another major danger. It causes temperatures and weather patterns to change faster than many species can adapt. Warmer temperatures, for example, can melt Arctic sea ice, threatening animals such as polar bears, which depend on it to hunt. Extreme weather events, such as hurricanes, lengthy droughts, and intense wildfires, can destroy habitats and wipe out entire populations of species. Climate change can also shift where plants grow, which affects animals that rely on those plants for food. "What may have been suitable habitat in the past may not be suitable habitat in the future,"[8] says Ted Weber, a policy analyst at Defenders of Wildlife. Species that cannot move or adapt quickly enough may face serious population declines.

Chemicals, plastics, and oil spills are all forms of pollution that harms animals and plants. Pesticides, fertilizers, and industrial waste can poison water and soil. These chemicals can weaken animals, reduce their reproductive ability, or cause birth defects. Bits of plastic float in oceans, rivers, and lakes, where animals may mistake the particles for food. Sea turtles, birds, and fish can become sick or die after swallowing plastic. Oil spills are especially dangerous, coating animals' fur or feathers and making it hard for them to stay warm or afloat. Pollution can spread far from its source, affecting ecosystems for many years.

Invasive species also threaten native wildlife by competing for food, spreading disease, or preying on native wildlife. Invasive species are plants or animals introduced into areas where they do not naturally occur. These species can spread quickly and cause serious problems. Without natural predators to keep them in check, invasive species can take over entire ecosystems. "Invasive alien species are a major threat to biodiversity and can cause irreversible damage to nature, including local and global species extinctions, and also threaten human wellbeing,"[9] says Helen Roy, an ecologist at the UK Centre for Ecology & Hydrology in Great Britain. Together, these threats push many species closer to extinction.

Conservation Challenges

Once a species becomes endangered, the path to recovery is often long and filled with challenges. One of the biggest challenges is funding. Protecting endangered species can be expensive. Conservation groups need funds to buy land to create protected reserves, restore habitats, pay scientists, enforce laws, and educate the public. Governments and organizations must choose how to allocate limited funds, and wildlife protection often competes with the needs of schools, health care, and infrastructure. In poorer

Oil saturates birds after oil spills, making it hard for them to eat, stay warm, or float. Birds that are not helped through human intervention often die.

regions, there may not be enough money or trained workers to monitor endangered species or stop illegal hunting. Without steady funding, even well-planned conservation projects can fail.

Another challenge arises when conservation projects conflict with the needs of human communities. For example, as a city grows, its residents need land for homes, roads, and businesses. This development can destroy or fragment the natural habitats where endangered species live. In some areas, residents rely on hunting, fishing, or logging to earn a living, making it difficult to limit these activities.

The complexity of ecosystems can also be a challenge. Ecosystems are made up of many living and nonliving parts, such as soil or water sources, that depend on one another. When one species disappears, it can affect many others. Restoring damaged ecosystems takes time and careful planning. Even if a species is protected, it may still struggle if its food sources, predators, or climate conditions have changed. Scientists often need many years to fully understand and repair these complex systems.

International cooperation is also required when endangered species and their habitats span multiple countries. For example, tigers are endangered, with approximately fifty-five hundred animals in the wild in 2025, according to the WWF. Tigers need large territories, and their habitats stretch across several countries, including Bhutan, China, India, Nepal, and Russia. If one country protects tigers but a neighboring country does not, conservation efforts may fail.

Identifying endangered species and the threats they face is the first step toward effective conservation. With this knowledge, conservationists can develop a recovery plan that includes steps such as habitat restoration, captive breeding programs, or pollution cleanup. "All these species, whether big or small, are essential elements of ecosystems. When they are lost or depleted, nature's careful balance is disrupted. We cannot bring back the passenger pigeon or the Carolina parakeet and other extinct species, but we have the power to save the endangered species that remain and restore the biodiversity fundamental to our own well-being,"[10] says Lauren McCain, former senior policy analyst at Defenders of Wildlife.

Chapter Two

The Power of Habitat Conservation

Habitat loss is one of the greatest threats to Earth's animal and plant species. Around the world, critical habitats are disappearing as land is cleared for human activities. Natural areas are destroyed, broken into smaller pieces, or damaged so badly that the animals and plants can no longer survive there. Human activities such as farming, oil and gas drilling, building cities and roads, and diverting water can significantly change ecosystems. When this happens, habitats may no longer provide enough food, clean water, shelter, or safe places for animals to raise their young. As a result, Earth's species have fewer places to live. According to the IUCN Red List, 85 percent of endangered and threatened species face habitat loss as a primary threat to their survival. Other environmentalists agree with this assessment. "Loss of habitat is one of the biggest drivers—in many cases the biggest driver—for species loss,"[11] says Shelley Luce, campaign director of the Sierra Club, an environmental organization.

> **"Loss of habitat is one of the biggest drivers—in many cases the biggest driver—for species loss."[11]**
>
> —Shelley Luce, campaign director of the Sierra Club

Establishing Protected Spaces

Saving endangered species often starts with habitat conservation, which means caring for natural environments so plants and animals can survive. Habitat conservation efforts may include protecting important habitat areas, restoring damaged ecosystems, and carefully managing land use to reduce harm to native species. One part of habitat conservation is establishing protected spaces, such as national parks, wildlife refuges, and sanctuaries. These spaces provide safe havens for vulnerable species and protect them from human activity and disturbances. Scientists and conservationists also use these protected spaces for research and educational efforts.

The grizzly bear is an endangered species that has benefited from protected habitats. Before 1800, an estimated fifty thousand grizzly bears roamed across the western portion of North America, particularly in the Rocky Mountains, along the Upper Missouri River, and in California. As European settlers pushed westward during the 1800s and established farms and settlements, the grizzly bear population was threatened by habitat loss and overhunting. By 1975, only between seven and eight hundred bears remained in the continental United States, primarily in Washington, Idaho, Montana, and Wyoming. In 1975, the grizzly bear population in the continental United States was listed as threatened under the ESA.

Since then, government agencies, conservation groups, and private citizens have worked together to save the grizzly bear. In 1993, the FWS created a grizzly bear recovery plan that identified six recovery ecosystems: the Greater Yellowstone Ecosystem, Northern Continental Divide Ecosystem, Cabinet-Yaak Ecosystem, Selkirk Ecosystem, North Cascades Ecosystem, and Bitterroot Ecosystem. In each area, the bear population would be managed, and its habitats would be protected to help the species recover. As of 2025, thanks to these efforts, the grizzly bear population in the lower forty-eight states has increased to more than nineteen hundred bears.

Protecting Panda Habitat

In China, the giant panda once lived throughout southern and eastern China as well as in Myanmar and northern Vietnam. The panda primarily lived in temperate mountain forests, where abundant bamboo provided food. However, clearing forests for farms

These pandas live in a panda sanctuary in China. The country has created a large network of panda reserves that protect more than 66 percent of giant pandas living in the wild.

and settlements led to habitat loss and fragmentation. The giant panda was slowly restricted to isolated patches of bamboo forest in China's Sichuan, Shaanxi, and Gansu provinces, preventing it from finding enough bamboo and potential mates. The combination of habitat loss and poaching led to a steep decline in the panda population. During the 1990s, the IUCN classified the giant panda as endangered.

Because habitat loss was the most serious threat to the panda, the Chinese government worked with conservationists to establish panda reserves. Currently, China has created a network of sixty-seven panda reserves that protect more than 66 percent of giant pandas living in the wild and nearly 54 percent of their remaining habitat. China has also worked with conservationists to develop "bamboo corridors," which connect isolated pockets of forest where pandas live. These corridors allow pandas to travel to new areas, meet new mates, and find additional food.

Slowly, the giant panda population has begun to recover. In 2016, after more than twenty years on the endangered species list, conservationists with the IUCN Red List reclassified

> "This iconic species, which is the poster child of endangered species globally, no longer qualifies as endangered."[12]
>
> —Ron Swaisgood, director of applied animal ecology at the San Diego Zoo Institute for Conservation Research

the panda as vulnerable. While threats to the panda species remain, the danger of extinction has lessened. "While we do not believe the giant panda is completely safe, our IUCN Red List evaluation highlights how far we have come in panda conservation. This iconic species, which is the poster child of endangered species globally, no longer qualifies as endangered. All the trends support this conclusion: Habitat is increasing, and the population is growing,"[12] said Ron Swaisgood, director of applied animal ecology at the San Diego Zoo Institute for Conservation Research, in 2016.

In 2024, China approved one of the world's largest protected areas to create a reserve for the giant panda on a total of 10,476 square miles (27,133 sq km). The protected area sprawls across three Chinese provinces and is three times the size of Yellowstone National Park. The massive reserve will connect the country's existing sixty-seven panda reserves on six separate mountain ranges. Chinese officials explained that they hoped that connecting the panda reserves would allow the animals a greater choice of mates, create a more diverse gene pool, and help increase the panda population. According to Hou Rong, director of the Chengdu Research Base of Giant Panda Breeding, the new reserve will be "a haven for biodiversity and provide protection for the whole ecological system."[13] This new national park will also be a space where eight thousand other endangered plants and animals receive protection.

Restoring Damaged Environments

Restoring damaged habitats is another important part of habitat conservation and protecting species. Natural disasters and human actions have left many ecosystems damaged and unable to provide the clean air, water, soil, shelter, and food that are essential for the species that live there. Habitat restoration is the process of returning damaged habitats and ecosystems to a healthy state. Restoring damaged areas allows native plants and animals to thrive and rebuild their populations. Some of the most common and effective habitat restoration methods include reforestation, wetland restoration, and invasive species removal.

In areas where forests have been cut down or damaged, reforestation—or the replanting of native trees—is a common restoration method. Native trees help prevent soil erosion and provide shelter for many animals. Replanting trees allows forests and the species that live there to grow back over time. For example, the endangered Carolina northern flying squirrel lives in high-elevation mixed-spruce forests of the southern Appalachian Mountains. These squirrels move through their forest by leaping from tree to tree. Habitat loss is one of the primary causes of the squirrel population's decline. However, efforts to plant new trees and restore the squirrel's habitat are helping the species recover. Since 2011, the National Forest Foundation has planted thousands of seedlings, including red spruce, table mountain pine, and northern red oak, in the central and southern Appalachian Mountains to restore forests damaged by human activities.

Saving the Galápagos Giant Tortoise

The Galápagos giant tortoise is one of the most recognizable animals on the Galápagos Islands, located about 600 miles (966 km) off Ecuador's coast. When human settlers arrived in the Galápagos Islands, they brought non-native animals, including goats, pigs, and rats. These animals, particularly the goats, ate large amounts of native plants, leaving little food for the giant tortoises. Their grazing also destroyed the tortoise's nesting areas. Rats and pigs ate tortoise eggs and hatchlings, and few young tortoises survived into adulthood. Because of human hunting and the invasive species, the tortoise population on the islands declined sharply. To support the endangered giant tortoise, conservationists worked with the Galápagos National Park to remove invasive species and restore native vegetation. Once these invasive animals were removed, native plants began to grow back, providing food and shelter for tortoises once again. Captive breeding and reintroduction programs also helped boost tortoise numbers. Today, several populations of Galápagos giant tortoises are increasing. Their recovery demonstrates the impact habitat restoration and invasive species removal can have on endangered species.

Restoring Wetlands

Wetland habitats support many plant and animal species and also help filter polluted water. Restoring wetlands often involves planting native plants and fixing water flow so the area can once again function naturally. The fen orchid is a rare wild orchid found in the United Kingdom, where it grows in fenlands (a type of wetland mainly fed by groundwater) and coastal sand dunes. For many years, this plant's population declined due to habitat loss, wetland drainage, and the establishment of farms in the fenlands. At one point, fewer than one thousand plants remained, and the species was classified as endangered on the Botanical Society of Britain and Ireland (BSBI) Red List.

Conservationists from several organizations began working together to restore the plant's habitats by raising water levels, clearing overgrown vegetation, and managing grazing so the orchid could get enough sunlight and space to grow. These efforts helped re-create the damp, open conditions the plant needs to survive. As a result, the fen orchid population had increased to

A fen orchid in a nature reserve in Wales. As a result of conservation efforts, this plant has been downgraded from an "endangered" to a "near threatened" classification.

more than ten thousand plants in 2025. Recognizing its recovery, the fen orchid was moved into a "near threatened" classification on the BSBI Red List in 2025. "The challenge now is to ensure that the Fen Orchid continues to thrive in its current locations and, either naturally or with a helping hand, flourish in former sites and new places within its native range in the coming years,"[14] says Peter Stroh, scientific officer at the BSBI.

Removing Invasive Species

In some cases, restoring natural habitats involves removing invasive species. Invasive species can spread quickly and push out native plants and animals. Removing these harmful species helps protect ecosystems and gives native plants and animals a better chance to survive and thrive. This strategy helped endangered island fox populations on California's Channel Islands recover. Island foxes live on six of the eight Channel Islands, located off the coast of Southern California, with each island home to a separate subspecies of fox. When Europeans settled on the Channel Islands, they brought animals that did not naturally belong there. Pigs, sheep, deer, feral cats, and other introduced species disrupted the balance of the island's ecosystem. These changes helped increase the number of golden eagles drawn to the islands, which began preying on island foxes. As a result, the island fox population had dropped by 90 percent by the late 1990s. Because of this serious decline, the IUCN listed the four island fox subspecies as critically endangered.

During the late 1990s, conservationists launched a recovery program to save the island fox. Among other efforts, the program included removing and relocating the golden eagles that hunted foxes, as well as removing the non-native feral pigs and other introduced grazing animals. These efforts, along with captive breeding and reintroduction of foxes, saved the island fox from extinction. The island fox population has since rebounded, and in August 2016, three fox subspecies were removed from the endangered species list. A fourth subspecies was upgraded from endangered to threatened.

Water Development and Freshwater Habitats

Freshwater habitats such as rivers, lakes, and wetlands are extremely important, but they are facing serious threats from human activities. These ecosystems provide most of the world's drinking water and support over 40 percent of all fish species, making them essential for both people and wildlife. However, freshwater environments are being damaged faster than land ecosystems. In recent decades, more than 20 percent of known freshwater fish species have become endangered or extinct. Water development, when humans build dams and canals, dig wells, and pump water for human use, threatens freshwater habitats by changing how water naturally flows through rivers, lakes, and wetlands. Dams and water diversions also affect water quality by stopping nutrients from moving downstream. Water development can also block fish from migrating to spawning areas, affecting their ability to reproduce. Also, large amounts of water withdrawn for farms, cities, and industry can lower water levels, which shrinks habitats and dries out wetlands. Under these conditions, native fish, amphibians, and plants find it more difficult to find food, shelter, and safe places to reproduce. Water development can also increase pollution because less water is available to dilute harmful chemicals and runoff from farms or urban areas. Together, these threats weaken freshwater ecosystems and reduce their ability to support wildlife.

Building Wildlife Corridors

Wildlife species can suffer when their habitat is fragmented, or divided into smaller pieces, by the development of farms, settlements, and roads. In rivers and lakes, dams and water diversions can block and divide aquatic habitats. Many animals need large areas to find food, water, potential mates, and shelter. When habitats are fragmented, the resulting smaller, disconnected areas may not be adequate for finding food and mates. Habitat fragmentation also makes it more difficult for migrating animals to find places to rest and eat on their migration routes.

One strategy to restore fragmented habitats is building wildlife corridors to connect separated areas and isolated populations of species. Wildlife corridors can be bridges, tunnels, or other protected paths that allow wildlife to travel freely. These corridors provide mammals, birds, and amphibians with safe paths between habitats, allowing them to avoid areas where they might encounter humans. When animals can move between habitats, they are more likely to find mates from other populations. This encourages genetic diversity, making species stronger and less likely to die out due to inherited diseases or harmful traits. Corridors also make it easier for wildlife to reach food, water, and shelter, especially when some areas are affected by droughts, wildfires, or seasonal changes. As climates change, animals may need to move to new areas to survive. Wildlife corridors make this movement possible.

Rebuilding the Florida Panther Population

The Florida panther is an endangered animal that has benefited from wildlife corridors. Historically, Florida panthers lived in warm-climate habitats, such as wetlands, swamps, and upland forests, stretching from Florida to Louisiana. However, development fractured their habitats, and hunting decimated the population. "They retreated to the southern tip of Florida in part because it was the last place they could go to hide from all the humans trying to hunt them,"[15] says Jennifer Korn, a panther biologist. By the 1970s, only about twenty to thirty Florida panthers remained living in the

An aerial view of a wildlife crossing in Germany. Corridors like this prevent wildlife from being struck by cars and other vehicles. The corridors allow animals to move between habitats so that they can find mates or food.

wild. It was one of the first species added to the US endangered species list after passage of the ESA in 1973.

To rebuild the panther population, during the 1990s conservationists brought in eight female panthers from Texas to breed with the Florida population. Slowly, the panther population increased. However, habitat fragmentation was still a problem. Like many wild animals, panthers need a lot of space in which to roam. A single male panther "needs 200 square miles of habitat, so the only way to save the species is multiple adjacent lands working together as one connected habitat,"[16] says Carlton Ward, a Florida conservationist and wildlife photographer. Speeding vehicles also posed a danger to panthers as they attempted to cross roads and highways.

"There is no recovery for the Florida panther without protection of the Florida Wildlife Corridor."[17]

—Carlton Ward, conservationist and wildlife photographer

To protect the endangered Florida panther and other wildlife, scientists and conservation groups planned important wildlife pathways across the state. Over time, their work became the Florida Wildlife Corridor, stretching across nearly 18 million acres (7.3 million ha) and extending from one end of the state to the other. The corridor allows wildlife to move safely across Florida to find food, mates, and shelter while reducing dangers such as road crossings. In 2021, Florida officially recognized and protected the corridor by passing the Florida Wildlife Corridor Act.

In 2016, researchers working at Florida's Babcock Ranch Preserve, located about 13 miles (21 km) northeast of Fort Myers on Florida's Gulf Coast, confirmed the first sighting of a female Florida panther north of the Caloosahatchee River in more than forty years. This finding gave strong support to the idea that wildlife will move into new areas when habitats are connected. "There is no recovery for the Florida panther without protection of the Florida Wildlife Corridor,"[17] Ward says.

Habitats worldwide provide the food, water, shelter, and space that animals and plants need to survive. Without suitable habitats, species struggle to survive. By restoring vital habitats, people worldwide can help endangered species recover and thrive.

CHAPTER THREE

Protection from Poaching and Overharvesting

Many of the world's largest and most awe-inspiring animals, such as whales, elephants, rhinos, and tigers, have become threatened because they were hunted too much in the past. People hunted whales for oil and meat, elephants for ivory, and rhinos and tigers for valuable body parts and as trophies. Even though laws now protect these animals, illegal hunting, called poaching, still occurs.

People take plants and animals for food and medicine, or they use them to make products such as wood furniture or leather shoes. Some plants and animals are also overharvested, meaning they are taken faster than they can replenish populations naturally. Overharvesting can cause populations to shrink and even disappear. For example, overfishing has caused the once-abundant codfish to become so scarce that they can no longer be caught in many areas along the US East Coast and in the North Sea. "Exploitation of wildlife represents one of the greatest threats to species survival,"[18] says Alice Hughes, conservation expert and University of Hong Kong associate professor.

"Exploitation of wildlife represents one of the greatest threats to species survival."[18]

—Alice Hughes, conservation expert and University of Hong Kong associate professor

Protected Under the ESA

Legal protection plays a significant role in stopping poaching and overharvesting, which are two of the biggest threats to endangered plants and animals. Without strong laws and their enforcement, many species would continue to decline and could disappear forever. In the United States, one of the most important conservation laws is the ESA. Passed in 1973, the ESA protects animal and plant species that are at risk of extinction. "Once a species is listed, it gets legal protection,"[19] explains Brenna Forester, a biologist with the FWS. Listing a species as endangered or threatened makes it illegal to hunt, harm, or trade that species without special permission. The law also protects critical habitats, such as breeding areas and feeding grounds, that these species need to survive.

The ESA's goal is to rebuild species populations to the point that they no longer need protection. Biologists with the FWS work with species experts and other government agencies to develop recovery plans for each species listed under the ESA. The recovery plan outlines the steps to restore a species to health. Through its legal protections and recovery plans, the ESA has helped save animals such as the bald eagle from extinction.

Another species that has benefited from the ESA is the American alligator. This reptile historically lived in wetlands in the southeastern United States. For years, people hunted alligators for their skin, which was used to make leather for boots, belts, purses, and other luxury goods. Eventually, demand for alligator leather exceeded the supply, and the alligator population began to decline. By the 1960s, overhunting and habitat loss put the alligator on the brink of extinction. Several states banned alligator hunting, and the American alligator was also protected under the ESA. The ESA listing banned alligator hunting nationwide and protected their critical habitats. These state and federal legal protections, along with captive breeding programs, enabled the alligator population to rebound. By 1987, the American alligator population had recovered enough that it was removed from the ESA's endangered species list. Currently, limited trade of alligator products is allowed. "American alligators are an iconic American species that

"American alligators are an iconic American species that have healthy wild populations due to our work done in collaboration with state and local partners. Together, we have created a management system that preserves the species and supports sustainable international trade."[20]

—Hila Levy, assistant director for international affairs at the FWS have healthy wild populations due to our work done in collaboration with state and local partners. Together, we have created a management system that preserves the species and supports sustainable international trade,"[20] says Hila Levy, assistant director for international affairs at the FWS.

Protecting Endangered Marine Mammals and Birds

The ESA, along with several other laws, protects endangered species in the United States. The Marine Mammal Protection Act (MMPA) was passed in 1972 following a significant decline in several species of marine mammals due to human activity. Under the MMPA, it is illegal to harm, hunt, capture, or kill marine mammals in US waters without special permission. Marine mammals include whales, dolphins, seals, sea lions, manatees, and sea otters. The MMPA also protects habitats where marine mammals live, such as oceans, coastlines, and rivers. The act's goal is to ensure these animals can survive and thrive in the wild.

The Marine Mammal Protection Act helps protect manatees and other marine mammals and their habitats from human actions that threaten them.

In the years since its passage, the MMPA and its protection for marine mammals have been an essential part of conservation successes in the United States. The MMPA has been a key part of the recovery of populations of California sea lions, Hawaiian monk seals, and Florida manatees. Since the MMPA was passed, no marine mammal species has gone extinct in US waters.

Another important US law is the Migratory Bird Treaty Act (MBTA). Passed in 1918, the MBTA is one of the oldest endangered species laws in the United States. It was passed when North American bird species were declining due to overhunting and the unregulated trade in bird feathers, which were frequently used in women's hats at the time. The act bans the killing, hunting, selling, or possessing of most native bird species without a permit.

Another federal law, the Bald and Golden Eagle Protection Act (BGEPA), protects the iconic bald eagle and a related species, the golden eagle. The act makes it illegal to kill, capture, sell, or disturb eagles, as well as to damage or collect their nests, eggs, or feathers, without special permission from the federal government. Overhunting and poisoning from the pesticide DDT during the mid-1900s had significantly reduced the bald eagle population. By the early 1960s, an estimated 417 breeding pairs of bald

State Protections

In addition to federal laws, some states have also passed legislation to protect endangered species. The Florida manatee is a gentle, slow-moving marine mammal that lives in the warm coastal waters, rivers, and springs of Florida. The manatee was heavily hunted for its meat and oil and also faced threats from human activities, particularly boat collisions and habitat loss. In addition to federal protections under the ESA and MMPA, the Florida manatee is protected by the Florida Manatee Sanctuary Act. This 1978 state law established all state waters as a manatee sanctuary. It allows the state to set boat speed limits, create manatee refuges, and impose strict penalties for harming manatees. Under these legal protections and other conservation efforts, the Florida manatee population has begun to recover. In 1991, the FWS estimated that fewer than 1,300 manatees lived in Florida's waters. By 2022, scientists with the US Marine Mammal Commission estimated approximately 9,970 manatees live off Florida's coasts.

eagles remained in the continental United States. However, the legal protections banned the taking or possession of eagles as well as their feathers, eggs, and nests, which helped eagle populations recover. By 2007, the bald eagle was removed from the ESA, but it remains protected under the MBTA and BGEPA. By protecting eagles and their nesting areas, these laws have helped bald eagles recover from near extinction in the United States. Today, the BGEPA helps ensure that both bald and golden eagles remain safe and continue to live in the wild.

International Agreement: CITES

On an international level, the Convention on International Trade in Endangered Species of Wild Fauna and Flora (CITES) is a conservation agreement among more than 180 countries. CITES regulates or bans the international trade in endangered plants and animals and their products, such as ivory, animal skins, and rare plants. CITES covers about thirty-five thousand plant and animal species, which are classified into two tiers. For species listed in the first tier, known as Appendix 1, international commercial trade is banned. For species listed in the second tier, Appendix 2, tightly regulated trade is allowed. These bans and regulations reduce the demand for endangered animals by cutting off legal markets and making it much harder for buyers to get wildlife products. The bans and regulations also increase the risks of buying and selling wildlife products and make it harder for poachers and illegal traders to find buyers. Together, these actions reduce demand by limiting access, increasing consequences, and educating the public, helping endangered species recover in the wild.

For the big leaf mahogany tree (*Swietenia macrophylla*), CITES regulation has helped the species survive. The tree, which grows in the South American rainforest, is prized for its deep reddish color and durability. For many years, the tree's wood has been used as a raw material in the manufacture of high-quality musical instruments and furniture. Due to strong demand, overharvesting

Mahogany seedlings in an Ecuadorian nursery. An international conservation agreement is helping this tree—prized for its wood—survive.

threatened the big leaf mahogany before it was protected under CITES. "The trade is being monitored and regulated closely,"[21] says Susan Lieberman, vice president for international policy at the Wildlife Conservation Society.

Enforcement Efforts

Despite laws that protect endangered species, poaching and the illegal wildlife trade still occur. Strong enforcement efforts have helped reduce these illegal activities and punish offenders. In the United States, the FWS is the primary agency responsible for enforcing federal conservation laws. The agency works with enforcement officers from NOAA, as well as local, state, tribal, and international enforcement officers, to prevent and investigate poaching, illegal trading of wildlife and plants, and habitat destruction.

FWS enforcement officers have worked on numerous cases to protect endangered animals in the United States. One case involved the Atlantic salmon, which was listed as endangered under the ESA in 2001 when its numbers dropped sharply due to overfishing. Fishing for endangered salmon was banned, but some people continued to catch them illegally. In 2009, the FWS and the Maine Warden Service conducted a secret investigation into the illegal capture and sale of Atlantic salmon from the Piscataquis River. Undercover officers identified a suspect and observed the person illegally snagging salmon at night as they swam upstream to spawn. The officers also confirmed the suspect knew the fish were protected. When the officers searched the suspect's home and vehicle, they found fish remains, which DNA tests proved were from Atlantic salmon. The suspect was charged with violating the ESA and other conservation laws and was sentenced to six months in federal prison and two years of federal probation.

Stopping Poachers in Africa

In Africa, the black rhinoceros was a popular target for European hunters and settlers. During the 1900s, overhunting threatened the black rhino's survival. Between 1960 and 1995, the black rhino population decreased by 98 percent, with fewer than twenty-five hundred animals living in the wild. Since then, conservation efforts across Africa have helped the black rhino population rebound to almost sixty-five hundred in 2025. However, the species remains

Tracking the Illegal Ivory Trade

In 2016, an anonymous tip led FWS officials to investigate two men traveling to the United States to attend a billiards expo. Officials learned that the men planned to buy pool cues made from African elephant ivory. At the time, the African elephant was listed as threatened and protected under the ESA, and there was a ban on most commercial trade of elephant ivory in the United States. A wildlife inspector at the Port of Newark monitored the men's social media activity and online forums to track their plans. The inspector worked with US Customs and Border Protection to determine when and where the men would attempt to leave the country. Agents tracked the men to Los Angeles International Airport, where they searched the men's checked luggage and found pool cues made from forty-four pieces of African elephant ivory. Officials stopped the men before they boarded their flight and arrested them for smuggling wildlife products. The men served six months in jail, followed by probation. The investigation also led officials to the pool cue maker who supplied the ivory and helped plan the illegal export. He was arrested and sentenced to four months of home confinement, followed by two years of probation, and fined $10,000.

classified as critically endangered, and more work is needed to ensure its long-term survival. One of the biggest threats to the black rhino's recovery is illegal poaching for its horns, which are then sold on the international market. Between 2008 and 2021, thousands of rhinos were poached in Africa, according to the WWF.

Several anti-poaching efforts are being used in Africa to protect the black rhino and other illegally hunted animals. In Kenya, the Kenya Wildlife Service has established several rhino sanctuaries across the country. One such sanctuary is the Loisaba Conservancy, located less than 200 miles (322 km) north of Nairobi, Kenya's capital city. At Loisaba, strong security measures protect the rhinos from poaching. Full-time rangers patrol the sanctuary on foot and in trucks, twenty-four hours a day, seven days a week. A unit of trained tracking dogs and handlers also patrol the sanctuary, and a surveillance airplane monitors Loisaba from the sky. Loisaba also uses technology to protect the rhinos. Drones fly above the sanctuary to track the animals and spot potential

Kenya Wildlife Service officials hold two black rhino horns that were discovered being smuggled on a cargo plane. Illegal poaching is a massive factor in the sharp decline of the rhino.

poachers. Software gathers and combines real-time data from ranger reports, individual rhino GPS tags, field sensors, and cameras to create a live, comprehensive view of the sanctuary and its inhabitants. "Software like EarthRanger helps us track the movement and behavior of animals, aids in coordinating ranger patrols, and allows us to monitor potential threats like poaching,"[22] says Saruni Yiankere, a security manager for Loisaba.

Technology is also being used to monitor and protect Bengal tigers in India. One technology in use is a camera system that detects tigers and poachers in real time. These cameras use artificial intelligence (AI) to spot unauthorized people moving through tiger habitats. The camera sends alerts and images in real time to a control center, which deploys forest rangers and anti-poaching operations. This smart system helps officials respond quickly to poaching threats. In the Similipal Tiger Reserve, rangers set up about 100 to 150 smart camera units. Over ten months in 2024, the AI-powered cameras led to ninety-six poacher arrests and eighty-six firearm seizures. Officials estimate that the AI cameras have reduced poaching by 80 percent. "Earlier, we didn't know when poachers entered. Now we have clear information about the area the poachers are in, so our people are prepared,"[23] says Samrat Gowda, deputy director of the Similipal Tiger Reserve.

"Earlier, we didn't know when poachers entered. Now we have clear information about the area the poachers are in, so our people are prepared."[23]

—Samrat Gowda, deputy director of the Similipal Tiger Reserve

Drones carrying high-resolution and thermal cameras are also being used in India to monitor and protect wildlife and deter poachers. Able to fly at night, the drones can quickly cover vast, hard-to-reach areas and work with human ranger patrols. When the drones detect wildlife, poachers, or fire, they alert officials, who can then respond appropriately.

Strong laws and enforcement are essential to saving endangered animals. Legislation protects plant and animal species by making it illegal to harm or sell them, limits global trade, and reduces incentives for black market traders. The threat of penalties and punishment for violators deters potential poachers and gives endangered species a chance to recover.

Captive Breeding and Reintroduction Programs

For many years, tens of thousands of black-footed ferrets lived in North America's grasslands. These small predators lived near large prairie dog colonies, took shelter in prairie dog burrows, and hunted prairie dogs as their primary food source. As European settlers spread across North America, they converted traditional grasslands into farms and ranches. Farmers and ranchers also poisoned large numbers of prairie dogs to remove them from their land.

The black-footed ferret suffered from the dramatic decline in prairie dogs and their colonies. By 1967, the black-footed ferret was listed as endangered. In 1980, the last known black-footed ferret died in captivity, and conservationists believed the species was extinct.

Then, in 1981, a ranch dog on a farm in Meeteetse, Wyoming, killed a black-footed ferret and brought it to his owners. A local taxidermist identified the animal, and an investigation revealed a small population of 130 ferrets on the Wyoming farm. However, disease struck this tiny population, and by 1987, only 18 ferrets remained. In a desperate attempt to save the species, scientists captured the remaining black-footed ferrets and launched a captive breeding and reintroduction program.

What Is Captive Breeding?

Captive breeding is used to help endangered animals and plants reproduce in protected, human-managed settings such as zoos, aquariums, botanical gardens, or wildlife centers. Captive breeding programs are often used when threats such as habitat loss, poaching, or disease make survival in the wild difficult. In a captive breeding program, scientists carefully select which individuals should breed to maintain the population's health and genetic diversity. Some animals breed naturally, but others may need help through methods such as artificial insemination. In these programs, females typically raise their young, much like they would in the wild.

> **"The world is facing an extinction crisis, and many people are doing the best they can to save at-risk species. Captive breeding and release will become an increasingly important tool to save species from extinction."[24]**
>
> —Dejan Stojanovic, a conservation biologist with Australian National University

The primary goal of captive breeding programs is to increase population numbers and protect species that are struggling to survive in the wild. Once the population becomes stable, some animals or plants may be reintroduced into their original habitats to help restore wild populations. Others may be kept in the breeding program to help continue to increase the number of individuals. These programs also help scientists study a species' health and behavior. "The world is facing an extinction crisis, and many people are doing the best they can to save at-risk species. Captive breeding and release will become an increasingly important tool to save species from extinction,"[24] says Dejan Stojanovic, a conservation biologist with the Australian National University.

Conservation breeding programs typically focus on one species or a small number of closely related species. Often, zoos and aquariums worldwide collaborate on breeding programs, with the goal of building healthy, stable populations that can reproduce on their own. Scientists carefully manage the animals and plants across multiple facilities to reduce the risk of species loss due to disease, natural disasters, or other unexpected events. Some programs serve as backup populations to protect a species' numbers and genetic diversity in case wild populations decline. Other programs focus on research about endangered species and help educate the public about them and the threats they face.

Reintroducing the Black-Footed Ferret

For the black-footed ferret, previous attempts to breed the animal in captivity during the 1970s had not produced any surviving young. This time, during the late 1980s, biologists with the Wyoming Game and Fish Department and the FWS brought in domestic ferret breeders and reproductive experts from the Association of Zoos and Aquariums to help them learn more about the ferret's reproductive system. Together, these wildlife professionals successfully bred the animals in captivity. Currently, there are six facilities participating in the black-footed ferret's captive breeding program, including the FWS's National Black-Footed Ferret Conservation Center in Colorado; the National Zoo's Smithsonian Conservation Biology Institute; and zoos in Arizona, Colorado, and Kentucky as well as Ontario, Canada.

In the 1980s scientists in the United States launched a captive breeding and reintroduction program to save the black-footed ferret from extinction. Since 1991, black-footed ferrets have been reintroduced into the wild in eight states and three countries.

After the success of the captive breeding program, conservationists tackled the question of how to reintroduce the captive-born ferrets into the wild. First, each ferret experienced at least thirty days of preconditioning at the National Black-Footed Ferret Conservation Center in Colorado. During this time, they lived in outdoor pens where they were exposed to natural burrow systems and live prairie dogs. The pens protected ferrets from predators while allowing them to gain experience preying on prairie dogs. They were also vaccinated against common disease threats, such as canine distemper virus and plague. Preconditioning increased the likelihood that the ferrets would survive once released into the wild.

Since 1991, black-footed ferrets have been reintroduced into the wild in eight states and three countries. Currently, approximately five hundred black-footed ferrets live in the wild, according to the WWF. Another two to three hundred ferrets are in captive breeding programs. Scientists and volunteers monitor the reintroduced ferrets and ensure their health and survival. "They are one of those species that's on the brink of extinction. And the work that I and so many others do is keeping them from blinking out. We don't want to let them go. We want to make sure that they're still here for future generations to experience,"[25] says Tina Jackson, the former black-footed ferret recovery coordinator for the FWS.

Captive Breeding Challenges

Captive breeding programs are an important conservation tool, but they also have several potential problems. One major issue is a lack of genetic diversity caused by small population sizes. In a small population, there are limited options for breeding pairs, which reduces genetic diversity. When animals are closely related, harmful traits and inherited diseases are more likely to be passed down to future generations. Low genetic diversity can also make a species less able to adapt to environmental changes, such as new diseases or climate shifts.

There is also the risk of failure when captive-bred animals are released into the wild. Animals raised in captivity may not have learned important survival skills, such as finding food, avoiding predators, or interacting with others of their species. Even with careful planning, many reintroduced animals do not survive long after their release. This means years of effort may not result in a lasting wild population.

Captive breeding programs are also very expensive. They require specialized facilities, trained staff, veterinary care, genetic testing, and long-term monitoring. For example, the California condor captive breeding and reintroduction program in the United States has cost more than $35 million since 1987, making it one of the most expensive captive breeding programs in the world. Funding these programs can take money away from other conservation efforts that may have a greater impact, such as habitat protection or anti-poaching programs.

Finally, captive breeding does little to address the root causes of species' endangerment in the first place. Problems like habitat destruction, pollution, and climate change often remain unresolved. If these threats remain, animals released from captivity may have few safe places to live. For captive breeding to be successful, it must be combined with efforts to protect and restore natural habitats.

The Importance of Genetic Diversity

Every living organism has a unique set of genes that hold the necessary instructions for life. An organism's genes determine traits such as size, color, strength, and disease resistance. When a species has high genetic diversity, individual organisms differ from each other genetically. When genetic diversity is low, many individuals share the same genes. Genetic diversity is important because it helps species survive and adapt to environmental changes. If a disease, climate change, or new predator appears, some individuals may have genes that help them survive. These survivors can then reproduce and pass on the helpful traits to offspring. Without enough genetic diversity, a single threat could wipe out an entire population. Low genetic diversity can also cause health problems. In small populations, related individuals may breed with each other, leading to inbreeding. Inbreeding increases the risk of genetic disorders, weakened immune systems, and lower survival rates in offspring. In conservation, genetic diversity is very important for endangered species. Protecting a wide range of genes helps populations stay healthy, reproduce successfully, and recover over time.

The California condor is one species being bred in a captive breeding program. However, this program has cost more than $35 million, making it one of the most expensive captive breeding programs in the world.

Saving the Hawaiian Crow

Efforts to save the Hawaiian crow illustrate the challenges of captive breeding and reintroduction programs. The Hawaiian crow is native to the Hawaiian Islands but has been extinct in the wild since 2002. Biologists successfully bred the crow in captivity, increasing its population from fewer than 20 to more than 110 birds. However, reintroducing the crows into the wild proved more challenging because the captive-bred birds lacked the survival skills needed in the wild. Several attempts to reintroduce the bird on Hawaii's Big Island during the 1990s and late 2010s were unsuccessful, with high fatality rates. Some birds died from malnutrition, and others became sick. Natural predators, such as hawks, killed many. The reintroduction program was paused, and the remaining birds were removed from the nature reserve and returned to captivity. Meanwhile, researchers studied what had gone wrong. "Conservation doesn't happen overnight," says Hannah Bailey, conservation program manager for the Hawaiian Endangered Birds Program at the San Diego Zoo Wildlife Alliance. "We're still learning, and so are the birds."[26]

> **"Conservation doesn't happen overnight. We're still learning."[26]**
>
> —Hannah Bailey, conservation program manager for the Hawaiian Endangered Birds Program at the San Diego Zoo Wildlife Alliance

In November 2024, researchers were ready to try again. This time, instead of releasing the crows on Hawaii's Big Island, researchers released a group of five crows on Maui, where they hoped the birds would be safer because of Maui's smaller population of predatory hawks. If the crows could establish a successful wild population on Maui, researchers believed they would be one step closer to reestablishing the Hawaiian crow on the Big Island. Before the release, a federal study determined that the introduction of crows on Maui would not have a significant environmental impact on the island; crows, or a very closely related species, had lived on Maui in the past.

The group of crows spent several weeks in outdoor aviaries on Maui before being released into a forest nature reserve. At first, researchers dropped off food for the crows to smooth their transition to living in the wild. They also fitted each crow with a tracking harness, which the researchers use to follow the crows' movements and outcomes. "We want to see them using the landscape well. We want to see survival,"[27] says Bailey. Researchers will closely monitor the crow population as the reintroduction effort continues.

Plant Conservation Programs

Endangered plants can also benefit from captive breeding programs, also known as off-site (*ex situ*) conservation programs. These programs grow endangered plant species in safe, managed spaces such as botanical gardens, where they are protected from predators, weather, and habitat loss. Seed banks store seeds from endangered plants so they can be germinated in the future. The overall goal of these plant conservation programs is to restore the endangered plants to their native habitats.

One drawback of these plant conservation programs, however, is that they frequently do little to address threats, such as habitat loss or invasive species. As a result, off-site conservation often has the best results when combined with other conservation actions, such as habitat restoration or invasive species removal, that protect species in their native environments.

Cloning Endangered Species

In December 2020, the first cloned black-footed ferret was born. A female named Elizabeth Ann was the first endangered US species successfully cloned. Scientists cloned Elizabeth Ann using preserved tissue from a wild black-footed ferret named Willa, who had died during the 1980s. The endangered black-footed ferret population had limited genetic diversity, making the animals more susceptible to disease and genetic abnormalities and reducing fertility. However, Willa's genes contained many unique variations that were not found in the current black-footed ferret population. Scientists attempted to clone her to reintroduce those genes into the ferret population and increase its genetic diversity. In 2023, scientists cloned two more ferrets using Willa's genes, females named Noreen and Antonia. "Genetic diversity is critical for resilience to environmental change," says Megan Owen, vice president of conservation science at the San Diego Zoo Wildlife Alliance. "It's basically the raw material of adaptive evolution." In 2024, Antonia gave birth to two surviving offspring, marking the first time a cloned endangered species in the United States gave birth in a conservation program.

Will Sullivan, "Scientists Clone Two Black-Footed Ferrets from Frozen Tissues in Conservation Effort," *Smithsonian Magazine,* April 22, 2024. www.smithsonianmag.com.

Rebuilding Hawaii's Rare Plant Populations

Across the islands of Hawaii, several plant conservation programs aim to save some of the world's rarest plants. Nearly 90 percent of Hawaii's native plants grow nowhere else on Earth. "We have almost half of all endangered plants in the United States. So, the level of rarity in Hawaii is unsurpassed anywhere else in the world,"[28] says Matt Keir, a botanist with Hawaii's Department of Land and Natural Resources.

Hawaii also has one of the highest plant extinction rates in the world due to a combination of factors, including habitat loss, invasive species, and climate change. To address this crisis, scientists and conservationists created the Plant Extinction Prevention Program (PEPP). This program focuses on saving Hawaii's rarest plants before they disappear forever. Conservationists primarily focus on more than 260 species that have fifty or fewer individuals remaining in the wild. "We're seeing the ecosystems collapse

A reintroduced Ka'ū silversword plant in Hawaii. Hawaii has one of the highest plant extinction rates in the world, prompting the creation of the Plant Extinction Prevention Program, which helps protect threatened plants in Hawaii.

and we're losing these rare endemic species, we want to prevent their extinction so they can continue to play their role in the ecosystems of Hawaii—we'd like to continue with our work to make sure that we're not losing these species,"[29] says Keir.

PEPP sends teams into remote forests to collect fruit, seeds, spores, and plant cuttings from the endangered plants. Some plant material is preserved in long-term storage at seed banks, and other material is used to cultivate new plants in specialized nurseries. These efforts ensure that even if something happens to the wild plants, the species will not be completely lost. "Our field crews go out and find rare species—field surveys and remote areas, we collect plants, and we bring them in and store them in seed banks, and while ecosystem and restoration are ongoing, we can hold these species until they're ready to return out into the wild,"[30] says Keir. Once plants are successfully grown in nurseries, conservationists carefully plant them in protected areas to establish new populations. Over time, scientists monitor the plants and track their growth and survival.

One example of PEPP's work is its efforts to save *Cyanea rivularis*, a rare flowering plant native to Hawaii. Like many Hawaiian plants, *Cyanea rivularis* faced serious threats from habitat loss and invasive species. At one point, only a few individual plants remained in the wild. PEPP collected seeds and cuttings, grew new plants in nurseries, and protected the remaining wild plants. Nursery-grown plants were later replanted into safe, protected habitats. "The goal is to create self-sustaining, healthy populations of species back into areas where they were once abundant,"[31] says PEPP technician Susan Deans.

"The goal is to create self-sustaining, healthy populations of species back into areas where they were once abundant."[31]

—Susan Deans, technician with Hawaii's Plant Extinction Prevention Program

Captive breeding is an important tool for saving endangered species when wild populations become too small to survive on their own. By carefully managing breeding and protecting genetic diversity, conservationists can raise healthy plants and animals in safe environments. These programs also support research and public education. When combined with habitat protection and strong conservation laws, captive breeding can help rebuild populations and give threatened species a second chance at survival.

CHAPTER FIVE

Cleaning Up Pollution

For endangered species, pollution can be especially dangerous. Lindsay Rosa, former vice president of conservation research and innovation at Defenders of Wildlife, explains:

> Many types of pollution—air, water and soil among others—are increasing, with negative impacts for nature. Greenhouse gas emissions, agricultural fertilizers, plastic waste, oil spills and many other sources of pollution have had strong negative effects on the soil, freshwater, oceans and atmosphere that wildlife and people depend on. . . . At the time of listing as "threatened" or "endangered" under the U.S. Endangered Species Act, more than 430 species were significantly impacted by pollution.[32]

Pollutants can cause illness, weaken immune systems, reduce fertility, and lead to birth defects. Many endangered species already have small populations, and even small losses can push them closer to extinction. In some cases, pollution disrupts entire food webs, which has a ripple effect on many species at once. Because of these wide-ranging effects, reducing pollution is essential for protecting endangered species.

Banning Harmful Chemicals

One of the most effective ways to reduce pollution is to ban chemicals that are known to harm wildlife. These bans prevent dangerous substances from entering ecosystems before they can harm the plants and animals living there. A well-known harmful pesticide is DDT, which was widely used during the mid-1900s to control insects, particularly mosquitoes. People sprayed DDT on the walls and ceilings of buildings, where it could remain for months, and farmers sprayed DDT on their fields. At first, DDT appeared successful because it helped reduce mosquito-borne diseases such as malaria and typhus and increased crop yields by killing harmful pests.

However, scientists began to notice that populations of many animals, birds, and fish were declining. By the 1960s, some scientists suspected that DDT was the cause. Initially, many people believed that the amount of DDT being used was not harmful. Yet what they did not consider was the fact that DDT did not break down easily in nature. Instead, it accumulated in the bodies of animals that were exposed to it. DDT levels became more concentrated as they moved up the food chain, a process called biomagnification. Birds of prey that were higher on the food chain, such as bald eagles, peregrine falcons, and ospreys, were especially affected by concentrated levels of the pesticide. DDT caused their eggshells to become thin and fragile, leading to widespread nesting failures. By the early 1970s, many bird populations were in sharp decline.

In response to growing scientific evidence and public concern about DDT's toxicity, the United States banned the pesticide in 1972. Many other countries followed suit. After the ban, DDT levels in the environment slowly decreased. Over time, bird of prey populations began to recover.

The bald eagle is an example of an endangered species that benefited from the DDT ban. When DDT washed into waterways, aquatic plants and fish absorbed it. Bald eagles ate the contaminated fish, and the concentrated chemicals disrupted the eagles' ability to produce strong eggshells. Thin eggshells frequently broke or did not hatch to produce young. Bald eagle numbers declined, and in 1973 the bird was placed on the ESA's list of endangered species. However, the ban on DDT, along with

A bald eagle looks down at her chick. Due to conservation efforts, in 2007 the bald eagle was removed from the list of threatened species.

other conservation efforts, has allowed the bald eagle population to rebound. In the absence of DDT, the birds started producing stronger eggs and raising more chicks. By 2007, the bald eagle population had recovered enough that it was removed from the ESA's list of threatened species.

Reducing Industrial Discharges

Industrial pollution is another major threat to wildlife. Along with releasing air pollution, factories, power plants, and mining operations can release chemicals, heavy metals, and waste products into rivers, lakes, and coastal waters. These pollutants can poison aquatic life and make water unsafe to drink for both animals and humans.

In the United States, one of the most important laws addressing industrial pollution is the Clean Water Act, passed in 1972. This law establishes limits on the amount of pollution industries are allowed to discharge into waterways. It also requires companies to treat wastewater before releasing it and gives the government authority to enforce pollution standards. Before the Clean Water Act, many rivers were heavily polluted. Some were so contaminated that fish could not survive, and others even caught fire due to oil and chemical waste. After the law was implemented, water quality began to improve in many areas.

Cleaning up industrial pollution has helped endangered species, including several species of freshwater mussels. Freshwater mussels play an essential role in keeping rivers and streams healthy by cleaning the water. As the mussels filter the water for plankton, their food source, they also remove pollutants and debris. A single mussel can filter up to 20 gallons (76 L) of water daily. However, pollution has threatened their survival. In North America, 66 percent of freshwater mussel species are at risk, according to the US Geological Survey.

In some areas, pollution reductions and cleanup efforts have helped some mussel populations stabilize and even recover. One area where recovery is underway is the Spring River, which flows through Kansas and Missouri. For decades, the region's lead and zinc mining operations polluted the river. Many native animals struggled to survive, including freshwater mussel species. "The fish community was greatly reduced," says James Whitney, an associate professor of biology at Pittsburg State University in Kansas. "There [were] basically no mussels at all."[33] During the 1980s, major environmental cleanup efforts began on the river. In 2019, scientists and students from Pittsburg State University began studying the impact of the river's pollution cleanup on its ecosystem. They found evidence that some species, including some freshwater mussel species, are recovering. In 2024, they found two endangered and threatened mussel species downstream from a previously polluted part of the river, another encouraging sign.

> "It just shows how complicated conservation and ecology can be in a river system and how quickly things can change. There's always new threats coming about."[34]
>
> —James Whitney, associate professor at Pittsburg State University

However, the scientists also noticed a troubling development. Mussels in historically cleaner parts of the river, upstream from industrial dumping sites, are starting to struggle. The cause is unknown, but researchers are investigating whether ammonia runoff from newly established poultry farms in the region is affecting the river and its ecosystem. "It just shows how complicated conservation and ecology can be in a river system and how quickly things can change. There's always new threats coming about,"[34] says Whitney.

Plastic Pollution

Every day, plastic waste pollutes the environment, flowing into oceans, rivers, and soil through litter, stormwater runoff, and waste disposal sites. The United States alone produces an estimated 46 million tons (42 million t) of plastic waste each year, and less than 9 percent of that plastic is recycled, according to the WWF. Because plastic does not degrade easily, it can remain in the environment for hundreds of years.

Noise Pollution

Another type of pollution, noise pollution, threatens southern resident killer whales (SRKWs), which live in the waters of the Pacific Northwest. Wild populations of these orcas declined significantly during the 1960s and 1970s due to the aquarium trade. However, the SRKW population, which is genetically different than other orca species, has struggled to recover. As of 2024, scientists estimate there are only seventy-four SRKWs left in the wild. Noise pollution is one factor inhibiting their recovery. These orcas rely on sound to communicate, hunt, and navigate. However, heavy boat traffic in the region, including cargo ships, ferries, and recreational vessels, creates constant underwater noise that interferes with these behaviors. When boat noise masks their calls, orcas struggle to locate salmon, their primary food source, leading to hunger and stress. Chronic noise exposure can also disrupt social bonds within groups of whales, known as pods, and inhibit reproduction. Scientists warn that continued noise increases could push this already endangered population closer to extinction. Reducing vessel speeds, creating quiet zones, and using quieter ships are important steps to protect the orcas and give them a better chance of survival and recovery.

Abandoned fishing gear is one of the deadliest types of plastic waste in the ocean and often kills sharks, whales, turtles, and other creatures that live there.

Plastic pollution can be deadly for wildlife. Animals may become tangled in plastic debris or mistake it for food. Sea turtles, for example, often eat plastic bags, mistaking them for jellyfish. The bags can block their digestive systems and lead to death. Smaller pieces of plastic, known as microplastics, can be ingested by fish and invertebrates. These particles may carry toxic chemicals and can move up the food chain, affecting many species. Nearly seven hundred endangered species, including the Hawaiian monk seal and the Pacific loggerhead sea turtle, eat plastic waste or get caught in it, according to the Center for Biological Diversity.

One of the deadliest sources of plastic waste in the oceans is lost or abandoned fishing gear, also known as ghost gear. According to the WWF, ghost gear accounts for as much as 10 percent of plastic waste in the ocean. Ghost gear traps marine animals, pollutes habitats, and contaminates the marine food web as it degrades. Nearly half (45 percent) of all marine mammals on the IUCN Red List are impacted by ghost gear, according to the WWF. Getting tangled in ghost gear kills hundreds of thousands of marine mammals and sea turtles each year. According to Marco Lambertini, director general of WWF International,

> While the consequences of plastic waste are finally starting to receive the attention they deserve, there's still too little awareness about the catastrophic harm caused by ghost gear. This needs to change urgently given that it's the most deadly form of marine plastic debris and that it can linger in our oceans for centuries, wreaking havoc like some kind of immortal menace: continuously and cruelly killing whales, dolphins, seals, seabirds, turtles and sharks, and damaging vital ocean habitats.[35]

Beach and coastal cleanup programs help remove plastic and debris from habitats before they cause harm. Individual volunteers, conservation groups, and governments around the world participate in cleanup efforts that protect wildlife and raise public awareness. In Mexico, a Conservation International program trains divers to search for and remove ghost gear pollution near the country's beaches. Gloria Acevedo dives with a group of women on the northern coast of Mexico's Yucatán Peninsula. Acevedo and her fellow divers, all of whom work in the commercial fishing industry, volunteer for the cleanup dives. "What we hope to do by removing ghost net equipment is protect marine life and recover the ecosystems damaged by the abandoned nets and lines,"[36] says Acevedo.

Cleaning Plastic Pollution to Save Green Sea Turtles

Green sea turtles are one example of a marine species threatened by plastic pollution in addition to overexploitation for their meat, eggs, and shells. Green sea turtles mainly eat jellyfish, seagrass, and algae, but floating plastic bags, balloons, and food wrappers can look like jellyfish or seaweed in the water. When turtles swallow these items, it can block their digestive system,

Coastal cleanup programs help remove trash from the ocean before it harms the creatures that live there. This boat collects plastic and debris from a river in Cambodia to prevent it from harming animals that live in the water and keep it from flowing downstream into the ocean.

cause internal injuries, or make them feel full even though they are not getting any nutrition, leading to starvation. Fishing lines, plastic rings, and discarded nets can entangle turtles, making it hard for them to swim, come to the surface to breathe, or escape predators. Plastic debris can also damage seagrass beds, which are the main feeding areas for green sea turtles, by smothering the beds or blocking sunlight. By the 1980s, the turtles faced extinction.

Decades of conservation efforts, including efforts to clean up plastic pollution, have helped the sea turtle population recover. Removing plastic waste from beaches and oceans has made habitats safer and healthier.

In 2025, the IUCN updated the green sea turtle's status from endangered to least concern. "The ongoing global recovery of the green turtle is a powerful example of what coordinated global conservation over decades can achieve to stabilize and even restore populations of long-lived marine species," says Roderic Mast, cochair of IUCN's Species Survival Commission Marine

Red Spruce Recovery

During the 1970s and 1980s, red spruce trees in New England forests suffered significant damage from acid rain, which is caused by air pollution from coal-burning power plants and factories. These installations released sulfur dioxide and nitrogen oxides into the air, which later fell as acid rain. When the acid rain fell on the forests, it washed nutrients such as calcium out of the soil and damaged the spruce needles. Acidic clouds and fog also inhibited photosynthesis for the trees. As a result, the red spruce trees weakened, and the population declined across parts of New England. Wildlife that depended on the trees, such as birds and small mammals, also suffered. Conservation efforts to clean up the pollution impacting the red spruce have helped the species recover. The 1990 Clean Air Act significantly reduced sulfur dioxide and nitrogen oxide emissions, leading to less acid rain. As air quality improved, soils slowly began to recover, and red spruce populations recovered in many areas. Other efforts, including reforestation, monitoring, and protection of red spruce habitats, have also helped the tree species begin to recover.

Turtle Specialist Group. "Such approaches must focus not only on turtles but on keeping their habitats healthy, and their ecological functions intact. Sea turtles cannot survive without healthy oceans and coasts, and humans can't either. Sustained conservation efforts are key to assuring that this recovery lasts."[37]

> "Sea turtles cannot survive without healthy oceans and coasts, and humans can't either. Sustained conservation efforts are key to assuring that this recovery lasts."[37]
>
> —Roderic Mast, cochair of IUCN's Species Survival Commission Marine Turtle Specialist Group

Despite the encouraging signs, experts warn that more conservation efforts are still needed to ensure the green turtle's survival. The turtles still face threats from habitat loss, marine pollution, fishing nets that are in active use, ghost gear, and climate change.

Reducing pollution can have a powerful impact in the fight to save endangered species. Clean water, air, soil, and oceans are the foundation of healthy ecosystems. If pollution contaminates the habitats where species live, other conservation efforts may fail. Often, pollution reduction is most effective when combined with other conservation strategies, such as habitat protection and restoration, sustainable resource use, and captive breeding programs. Together, these conservation efforts can help endangered species survive and recover.

While conservation efforts have helped save many endangered species from extinction, more work remains to protect wildlife habitats and prevent species from being lost forever. "We cannot overstate how important biodiversity is to our everyday lives, yet we are losing it more rapidly than we have ever seen. A greater understanding of these threats to species and where they are having the greatest impact on our communities will help us stem the crisis at hand with strategic investments and conservation action,"[38] says Rosa.

"We cannot overstate how important biodiversity is to our everyday lives, yet we are losing it more rapidly than we have ever seen. A greater understanding of these threats to species and where they are having the greatest impact on our communities will help us stem the crisis at hand with strategic investments and conservation action."[38]

—Lindsay Rosa, former vice president of conservation research and innovation at Defenders of Wildlife

SOURCE NOTES

Introduction: Saving the Gray Wolf

1. Quoted in Natalie Brand and Anna Grazulis, "How a Smithsonian Lab Is Helping Threatened Species Get Off the Endangered List," *CBS Saturday Morning,* August 23, 2025. www.cbsnews.com.
2. Quoted in Brand and Grazulis, "How a Smithsonian Lab Is Helping Threatened Species Get Off the Endangered List."
3. Quoted in Harley Coldiron, "Problem Solvers: Trump Administration Pushes to Remove Gray Wolves from Endangered Species Act Protections," KTVZ 21, December 11, 2025. www.ktvz.com.

Chapter One: The State of Endangered Species

4. Quoted in Shubhangi Dua, "World's Smallest Elephant Heads Toward Extinction, Just 1,000 Left in Wild," Interesting Engineering, June 28, 2024. https://interestingengineering.com.
5. Quoted in Michael Phillis, "Endangered Species List Grows by 2,000. Climate Change Is Part of the Problem," Associated Press, December 11, 2023. https://apnews.com.
6. Ellen Richmond, "Defenders Testifies on the Importance of the Endangered Species Act," Defenders of Wildlife, July 9, 2024. https://defenders.org.
7. Richmond, "Defenders Testifies on the Importance of the Endangered Species Act."
8. Quoted in Yale Climate Connections, "Climate Change Threatens Habitats That Many Endangered Species Need to Thrive," February 17, 2025. https://yaleclimateconnections.org.
9. Quoted in Intergovernmental Science-Policy Platform on Biodiversity and Ecosystem Services, "Media Release: IPBES Invasive Alien Species Assessment," September 4, 2023. www.ipbes.net.
10. Quoted in Glenda S. Booth, "Why Save Endangered Species?," Virginia Department of Wildlife Resources, December 6, 2023. https://dwr.virginia.gov.

Chapter Two: The Power of Habitat Conservation

11. Quoted in Benjamin Shingler, "World's Globetrotting Animals at Risk Due to Habitat Loss, Climate Change," CBC News, February 12, 2024. www.cbc.ca.

12. Quoted in Rafael Avitabile, "Giant Pandas No Longer Endangered, Still Considered 'Vulnerable,'" NBC San Diego, September 4, 2016. www.nbcsandiego.com.
13. Quoted in One Earth, "China Creates Giant Panda Heaven," July 10, 2024. www.oneearth.org.
14. Quoted in Neve Gordon-Farleigh, "Rare Orchid Brought Back from Brink of Extinction," BBC News, November 9, 2025. www.bbc.com.
15. Quoted in Isaac Eger, "Connecting the Florida Wildlife Corridor," National Wildlife Federation, September 27, 2023. www.nwf.org.
16. Quoted in Eger, "Connecting the Florida Wildlife Corridor."
17. Quoted in Eger, "Connecting the Florida Wildlife Corridor."

Chapter Three: Protection from Poaching and Overharvesting

18. Quoted in Florian Steinkröger, "New Study: Conservation Experts Warn of Current Dangers Posed by the Legal Wildlife Trade," Phys.org, May 11, 2023. https://phys.org.
19. Quoted in Jaime Chambers, "Explainer: What Is an Endangered Species?," Science News Explores, March 10, 2022. www.snexplores.org.
20. Quoted in Olivia Loyack, "Gator Gold: How American Alligator Permits Benefit the U.S. Economy," US Fish & Wildlife Service. www.fws.gov.
21. Quoted in Science Friday, "50 Years Later, Reflecting on the Treaty That Controls Wildlife Trade," March 17, 2023. www.sciencefriday.com.
22. Quoted in Jennifer Winger, "Return of the Rhino," The Nature Conservancy, May 8, 2025. www.nature.org.
23. Quoted in Nikhil Sreekandan, "AI-Fit Cameras in Similipal Tiger Reserve Send Poaching Plummeting," *The Hindu,* February 17, 2025. www.thehindu.com.

Chapter Four: Captive Breeding and Reintroduction Programs

24. Quoted in Australian National University Reporter, "Captive Breeding Impacts Chances of Survival in the Wild," November 21, 2022. https://reporter.anu.edu.au.
25. Quoted in Olivia Young, "Colorado Ferret Expert Now Out of a Job Due to DOGE Cuts Worries for the Future of a Species She's Dedicated Her Life To," CBS News, March 10, 2025. www.cbsnews.com.

26. Quoted in Elizabeth Anne Brown, "Hawaiian 'Soul Guide' Crow Returns to the Wild in Conservation Win," *Scientific American,* December 4, 2024. www.scientificamerican.com.
27. Quoted in Brown, "Hawaiian 'Soul Guide' Crow Returns to the Wild in Conservation Win."
28. Quoted in Casey Lund, "Hawaii Program Among 10 Recognized Nationally for Extinction Prevention Efforts," Hawaii News Now, December 6, 2023. www.hawaiinewsnow.com.
29. Quoted in Eric Natkin, "Preservation Efforts Involving Hawaii's Rare Plants Remains a Dilemma," Island News, December 29, 2023. www.kitv.com.
30. Quoted in Natkin, "Preservation Efforts Involving Hawaii's Rare Plants Remains a Dilemma."
31. Quoted in Hawaii Department of Land and Natural Resources, "One Beautiful Plant Highlights Endangered Species Recovery Efforts," December 6, 2023. https://dlnr.hawaii.gov.

Chapter Five: Cleaning Up Pollution

32. Lindsay Rosa, "Pollution, One of the Five Drivers of Biodiversity Loss," Defenders of Wildlife, December 7, 2022. https://defenders.org.
33. Quoted in Celia Llopis-Jepsen, "A Long-Polluted Ozark River Appears to Finally Be Bouncing Back from Disaster," Kansas City Public Radio, December 10, 2025. www.kcur.org.
34. Quoted in Llopis-Jepsen, "A Long-Polluted Ozark River Appears to Finally Be Bouncing Back from Disaster."
35. Quoted in World Wide Fund for Nature, "New Report from WWF Says Abandoned Fishing Gear an 'Immortal Menace' Which Must Be Central in the Fight Against Plastic Pollution." https://wwfwhales.org.
36. Quoted in Olivia Ferrari, "'Ghost Gear,' or Abandoned Fishing Equipment, Is Haunting the Oceans. Here's How Conservationists Are Fighting Back," *Smithsonian Magazine,* October 31, 2025. www.smithsonianmag.com.
37. Quoted in Rob Hutchins, "Green Sea Turtle Saved from Extinction in Major Conservation Victory," *Oceanographic,* October 10, 2025. https://oceanographicmagazine.com.
38. Quoted in Defenders of Wildlife, "New Defenders Report Shows Locations in the United States Most Impacted by Biodiversity Crisis," November 7, 2023. https://defenders.org.

FOR FURTHER RESEARCH

Books

Sneed B. Collard, *Defending Nature: How the US Military Protects Threatened and Endangered Species*. Millbrook, 2025.

Carol Hand, *Marine Conservation*. Abdo, 2025.

Tom Jackson, *Endangered Animals Atlas: A Journey Across the World and into the Wild*. Quarto Group, 2023.

Darcy Pattison, *Diego, the Galápagos Giant Tortoise: Saving a Species from Extinction*. Mims House, 2022.

Laura Perdew, *Conservation Success: Protecting Wild Spaces and Species*. Nomad, 2025.

Tracy Sue Walker, *Wildlife Conservation Technology*. Lerner, 2024.

Internet Sources

Natalie Brand and Anna Grazulis, "How a Smithsonian Lab Is Helping Threatened Species Get Off the Endangered List," *CBS Saturday Morning,* August 23, 2025. www.cbsnews.com.

Green Earth, "7 Strategies for Protecting Wildlife," April 7, 2023. www.green.earth.

Sophia Kagan, "How Endangered Species Have Fared in 2025—and What New Ones Made the List," WGAL 8, December 25, 2025. www.wgal.com.

Mary Kate McCoy, "Axolotls Are on the Brink. Can We Bring Them Back?," Conservation International, August 12, 2025. www.conservation.org.

Richard C. Paddock, Zunaira Saieed, and Sanjit Das, "In a Remote Jungle, the Gruesome Poaching of Rare Elephants," *New York Times*, December 16, 2025. www.nytimes.com.

Alexa Robles-Gil, "Sharks and Rays Gain Sweeping Protections from Wildlife Trade," *New York Times*, December 2, 2025. www.nytimes.com.

Organizations and Websites

Endangered Species Coalition

www.endangered.org

The Endangered Species Coalition works to protect wildlife by supporting strong environmental laws. It focuses on education and public action to defend endangered species in the United States. Its website has fact sheets, action guides, and other information about endangered species laws.

International Union for Conservation of Nature (IUCN)

https://iucn.org

The IUCN helps scientists track how close species are to extinction through the IUCN Red List. This information helps governments and conservation groups decide how to protect wildlife. Its website has information about species status, conservation data, and global biodiversity reports.

The Nature Conservancy

www.nature.org

The Nature Conservancy helps endangered species by protecting important land and water habitats. It works with local communities, businesses, and governments to care for natural areas. Its website offers maps, case studies, and explanations of how habitat protection works.

Wildlife Conservation Society (WCS)

www.wcs.org

The WCS protects endangered species through scientific research and conservation projects. It works in many countries to protect animals and their habitats. The WCS also helps educate the public through zoos and wildlife programs. Its website has articles, videos, and real-world examples of conservation science in action.

World Wildlife Fund (WWF)

www.worldwildlife.org

The WWF works around the world to protect endangered animals and the places they live. The WWF works with communities and governments to find long-term solutions that help both people and wildlife. Its website features species profiles, videos, maps, and simple explanations of conservation issues.

INDEX

PICTURE CREDITS

Cover: iStock

6: Michael Alden/Alamy Images
10: Avalon.red/Alamy Images
13: Alaskagirl8821/Shutterstock
15: Louisiana Governors Office/ZUMA Press/Newscom
19: Danita Delimont/Alamy Images
22: graham bell travel/Alamy Images
25: Arterra Picture Library/Alamy Images
29: Yelpar/Shutterstock
32: Morley/Alamy Images
34: Associated Press
38: Rick & Nora Bowers/Alamy Images
41: Georgi Baird/Shutterstock
44: Natural History Library/Alamy
48: John C. Beverly/Shutterstock
51: imageBROKER/Norbert Probst/Newscom
53: Associated Press

ABOUT THE AUTHOR

Carla Mooney is the author of many books for young adults and children. She lives in Pittsburgh, Pennsylvania, with her husband and three children.